What the F#*K is the Cloud?

James Bomford

First published by Busybird Publishing 2017

Copyright © 2017 James Bomford

ISBN
Print: 978-1-925585-34-6
Ebook: 978-1-925585-35-3

Cover image: Gabrijela Sklepic
Cover design: Busybird Publishing
Layout and typesetting: Busybird Publishing
Editor: Tom O'Connell

Busybird Publishing
PO Box 855
Eltham, Victoria
Australia 3095
www.busybird.com.au

Dedicated to Isabella, Charlotte and Freddie.

Living with an entrepreneur husband and father is not easy.

Thank you for your support!

CONTENTS

INTRODUCTION

S o what the fuck is the cloud?

I can tell you what it's not. It's not actual clouds, and no, storms do not affect access to your files. People often tell me they don't understand how it gets up there, and they ask "What happens when it rains?" I can say for certain that it's not actual clouds. I can also tell you that it's not just there. It's still in a physical location; you just don't know where that location is.

Cloud has been around for many years. I've been in the IT industry for about fifteen years, maybe a bit more if you count secondary school. What I have found is that the cloud has been around for a long time; it was just never called 'cloud'.

When IT people drew diagrams for management, the internet was commonly drawn as a cloud, with the various systems within the business connected to it. Clever marketing people came up with the idea to call any system that wasn't in a business the cloud.

There are various types of cloud computing. There's software as a service, infrastructure as a service and platform as a service.

'Software as a service' is a collection of systems that you're probably used to and have heard of before. Systems like Gmail, Hotmail, and accounting systems like Xero and Saasu. These systems run from a web interface, so you don't actually have to install any software on your machine to be able to run them. Various systems integrate both, like Evernote, a note-taking program that has evolved to handle handwritten, typed and photographed notes and does software as a service. You can log onto your Evernote account and access your notes, or download their software onto your computer and synchronise your notes with your machine.

'Platform as a service' is a set of tools and services designed to make coding and deployment of those applications efficient. The end user (you) never knows about these, nor do they really need to know. It's simply behind the scenes.

The next part is 'infrastructure as a service'. Those systems, like Xero and Gmail, Hotmail, etc., all run off infrastructure as a service. This means the actual systems running the software are in data centres managed by Microsoft and Google or other providers like Amazon, etc.

Some businesses actually have setup their business to use infrastructure as a service. For example, a business may require you to install their software to use it. In this case, an entire virtual server is setup and you can install your software onto that server. It's private, meaning no one else can access it except you, but you don't have to pay for the physical hardware or fix dead hard drives.

The benefit of having infrastructure as a service is that you can continue using the software that you're used to using, but also enjoy the convenience of a cloud-based server. The benefit of using software as a service is that you don't have to maintain the software, the server or the infrastructure.

These are all managed by the provider.

In short, the cloud is not some mystical thing that's in the air where your data is in the atmosphere. It's a simple term that just means something else is looking after everything for you, whether it be Google and Gmail, Xero and your accounting, Dropbox and your files or iCloud and your backups. Each of those systems will physically exist somewhere and you can have access to them for a monthly fee.

CHAPTER ONE

DEMYSTIFYING
THE CLOUD

W hat would you do if each of your employees had an extra 10 hours per week to complete their work?

Most business owners wouldn't believe that was possible. However, with cloud computing there are many solutions that can streamline business processes and systems, allowing employees to spend more time making the business money and less time performing administrative tasks.

From an IT perspective, the biggest costs to a business are usually ongoing maintenances, hardware replacements and the massive downtime that occurs when something goes wrong. There are also soft costs, which include hardware failures, meaning employees are unable to complete their jobs. Moving various systems to the cloud removes many of these soft costs from the business. These costs include electricity and employee downtime. Some businesses can save up to ninety-five percent on their electricity bill alone just by moving their systems to the cloud and retiring their server.

Other reductions in costs are downtime issues, which occur when the server is not working correctly and staff are sitting around twiddling their thumbs. For a small business with five employees, a server that's offline one hour per week can cost tens of thousands of dollars per year in lost wages and productivity.

I am often asked, "Is the cloud safe and secure?" The answer is that it's safer and more secure than actually having a server at your premises. The amount of money that goes into data centres to protect their servers is much more than any small business can afford. This is why the cloud is so secure and why they have a monthly fee.

Cloud computing is so successful because the businesses in charge of creating cloud servers and giving access to cloud providers can put a good part of the budget into ensuring that the servers and hardware are physically and virtually secure. Their data centres require very high levels of clearance to access. They also use military-grade software security, including firewalls, encryption and multiple failover options, which no small or medium-sized business would be able to afford.

Small businesses typically have their servers located at their premises. This makes them more susceptible to theft, hardware failure and flood or fire damage than a data centre. There will also be multiple data centres synchronising your data at all times. If one data centre has an issue multiple other servers will be ready to take the slack of the failed server. Hard drives and other hardware within the servers can literally be replaced while the servers are still running.

How can we get your employees to save 10 hours per week each? The answer is with the various cloud-based systems that interconnect to each other. You can have your customer relationship manager interconnected with your general ledger and finance systems, which themselves are connected to your online store and e-commerce solution. Each of these seamlessly connects, enabling a streamlined workflow process from the entry-level sales to the logistics process to the invoicing and payment scenario. Many businesses have cut their accounting expenses in half by taking advantage of these streamlined processes, so all that's needed is a simple bookkeeping exercise to ensure that all outgoing invoices match with the payments.

Logistics is also quite a large component of cloud computing. With it, you can use your system to track parcels or packages from freight companies. Being able to provide this information to clients and customers will improve your customer service and provide transparency across all of the systems for your clients, which in turn creates better relationships.

It's possible to have various systems in the cloud. For instance, you might want your accounting system, customer relationship manager, e-commerce solution and website to talk to each other. The best thing about the cloud is the API (Application Programming Interface), an application integration component that connects certain systems with other systems. Many of them already have this functionality. For example, an accounting system like Xero can connect directly to WooCommerce, which is an online eCommerce system.

Xero also has the same functionality to connect directly into E-Way, which is a payment gateway service for people paying for various invoices, etc., on credit card. The feeds from those credit cards will come directly into your accounting system and match automatically to the provided invoices, ensuring a streamlined process for the bookkeeper and the accounting staff.

The possibilities are certainly endless. Other companies out there have created cloud-based systems that help connect all your systems using triggers. Say, for example, someone buys something on your e-commerce website; you'll need an automated workflow that allows the product to be shipped once payment has been confirmed. A cloud-based trigger system can detect that a payment has been made and will send the required information to your logistics team so that they can have the product packed and shipped. Once the product has been shipped, the trigger system can then send out an automated email to advise the customer of the pending package arrival. The accounting team is also notified; however, as the payment and system invoice have already been connected, all they need to do is confirm the bank reconciliation.

The website, Zapier.com, enables multiple connections to multiple systems with multiple triggers. This means you can have five to six triggers coming from one system. Someone buys something on your website, which triggers the marketing list for that particular item; you can then send them an update whenever a new item becomes available. It also notifies the accounting team that something has been purchased and prompts the accounting system to create the invoice and match the payment.

These sorts of systems enable very small businesses to appear very large and streamline very large businesses to minimise administrative tasks. It's all about maximising profit and customer satisfaction, which in turn, increases the bottom line.

In contrast to all of this, you don't want to be left behind. Many businesses are moving ahead in leaps and bounds by implementing these sorts of systems. They have less staff doing more work, are making more sales, performing less administrative tasks and are improving customer satisfaction, whereas businesses with on-site systems can have many issues crop up. What if the internet does go down and you have all these systems in place? The best thing is that all of these cloud providers out there have multiple internet connections with multiple service providers in multiple areas. Therefore, the failover options are in place so that if one system goes down, another is already running. This means that work can continue as if nothing has changed; if one internet connection goes down, the other internet connections keep working. You won't see any difference in performance or in the way you use the system. It's like a duck floating on water. It all seems smooth and majestic, but the legs under the water are working hard to keep the duck moving forward. In this case, the cloud provider is the one working hard to ensure your experience is as smooth and worry-free as possible.

What happens when your internet goes down and all your systems are on premises? Your business communication and customer communication can stop, and if you're working remotely it has an even bigger impact. Not only would email stop flowing to your business, meaning all orders and employee communication stops, but if you're working remotely, you would have no access to any systems at all. This would mean lost revenue, lost sales and angry customers.

A cloud-based email system means that if your internet goes down, it's no big deal. Many modem routers come with the option to have a 2G, 3G or 4G USB connected or even a SIM card directly inserted into them. Additionally, many telecommunication providers enable twelve-month or twenty-four-month prepaid devices and accounts so that you don't have to pay ongoing costs for a redundant internet connection. This only kicks in when the internet goes down, and since customers aren't relying on your business having internet access (because they're not connecting into the systems that are in your business), your employees can work through that 2G, 3G, 4G internet connection, enabling a very smooth transition and continued customer satisfaction.

There are so many reasons to move to the cloud, but three key reasons are: reliability, security and mobility. I haven't touched much on mobility yet, but the systems I've mentioned will be especially beneficial for those wanting to break free from the office. You want to move around or work from a café. Sometimes you just need the freedom to work from anywhere, including while you're on holidays. I've had one scenario where a business was picketed and the employees couldn't actually enter their premises. Rather than having to fight their way through the crowds, they were able to go home, find another office space or visit a café with internet access to continue to work remotely using their iPads, other tablets, laptops or even their smartphones.

Admittedly it can be difficult to print because you don't always have that mobile printer, but the flexibility of not having to go into the office is ideal. Plenty of companies, like Office Works and Snap Printing, enable you to print documents by sending them via email or handing over a USB. We had a client turn up to work one morning to discover their entire building had burnt down. There was nothing left; all the employees were standing around looking at each other without knowing what to do. Rather than shutting down the business and possibly having to declare bankruptcy, the workers were able to go to a café and use the free Wi-Fi to continue working using their laptops, tablets and smartphones. They didn't even need to load software onto their devices, which is priceless for a business.

Obviously this outcome is not always possible. If you carry product and that product is destroyed, it does make it a little more difficult as you still need stock to make sales. However, if technology and communications aren't the cause of failure then you'll be able to focus on getting your stock back faster, which, in turn, will save the business tens of thousands – if not *millions* – of dollars.

I am constantly asked, "What if the cloud provider is hacked?" Here are a couple of things to be aware of. To begin with, it's more likely that you'll have your on-site server hacked than the cloud provider's. This is because adding/installing the amount of failover options and security layers that a cloud provider puts in place would again exceed what most small businesses can afford. Saying that, however, we've all heard incidents where large corporations face security breaches and have data stolen. Each time something like this happens, systems and security around the globe are improved. The plan is to ensure whatever went wrong never happens again. It's like the aviation industry. If something happens to a plane, they make extra sure that the issue is fixed on every other plane to prevent it from ever happening again.

The benefit to larger corporations having these sorts of issues is that it forces them to re-evaluate and look for better ways to secure their data. What ways can they protect your information? What other systems can they put in place? Cloud providers then benefit from this. They then implement extra security and take on all of this advice from these larger companies to ensure that your data is safe.

Other questions I am constantly asked are, "What if the cloud goes down?" and "What if the system you're using stops working?" Most of the time the redundancy options put in place by the cloud provider mean you will enjoy 99.99% uptime. So, yes, while the cloud systems do go down, these cloud providers have the technical expertise on hand to ensure that they come back up in a timely manner. This is because they're giving you the guarantee of the service level agreement, the legal obligation that they're going to keep their systems running for 99.99% of the time.

While the cloud does go down, or your systems do go down, this happens far less often than if you were running these systems on premises.

The other question I'm asked, which I've mentioned in this chapter, is "What if the internet goes down? What if we don't have internet at the office?" I've experienced scenarios where a business's internet and phone lines were cut because of disruptive road works in the area. The optical fibre was cut and we were told it was going to take days – if not *weeks* – to repair. Sometimes saboteurs go through certain telecommunications areas with an axe or some sort of an implement and completely shatter the environment. When this happens, it takes many technicians to re-cable, rewire and repair all the damage that's been done. In these cases phones go down, the internet goes down and there are a multitude of other issues. The thing is, you only need an internet connection to access your cloud system, whereas you'd need an internet connection and a working server at your office if remote employees are to access your on-premises server.

If your business has a router with a USB connection or SIM card option for a prepaid wireless 3G or 4G network, it will automatically kick over to this backup connection in the event that your hard line internet connection goes down. This means you can continue working.

If push comes to shove, there are cafés you can go to. You could try McDonald's, your neighbour's place, various business centres, or many other options. You just need to find an internet connection.

Chapter Two

SUNSHINE BEHIND THE CLOUD

How would you feel if you never had to think about IT again?

Many business owners that I talk to are always having IT issues. Either the server isn't working or the applications are crashing, leaving employees unable to do their job. The integration between systems is not working; therefore, it is costing the business more time and money because of the extra tedious administration tasks that staff now has to do.

Cloud systems enable freedom from the worries that come with information technology. Someone else is looking after the backend of the systems. Therefore, you don't have to worry about server maintenance that once upon a time plagued business owners. Although cloud computing is a foreign concept to some, it has fantastic benefits for businesses that are not usually considered. I'll touch on many of these to show how you can save – and even *make* – money by moving to cloud systems. Plenty of IT people will talk about how good a cloud system is, and how it integrates two systems within the business. But they don't talk about the actual benefits of the system and how the connection of these two systems will save you time and money.

As a business owner myself, I know that removing stress can greatly benefit a business, and cloud systems achieve this. You don't have to stress about downtime, loss of data, lack of connection or access. You can even eliminate the stress of entering the same data into multiple systems by integrating a trigger system that will automatically update other systems based on data you've entered before. Another business benefit is that the administrator knows exactly what their customers are buying in real time. This allows them to market directly to that customer based on what they are looking for, rather than throwing information on all your products and hoping that one stands out.

Cloud allows for mobility. This is the ability to access systems pretty much from anywhere in the world. I have had various clients running webinars and selling products online, a perfect example of mobility. One such client earned over $60,000 while sitting on a beach in Greece! The online ability to run an entire webinar and sell products with an online credit card facility means that products and services can be sold even during a live webinar. This is extremely powerful for businesses that want to streamline their online sales.

When working remotely you also get peace of mind knowing that your data is securely backed up. From an IT point of view, this is a far less stressful outlook. Employees are happier because everything works efficiently, and customers are happier because they get everything they want – usually in a faster and clearer manner.

I'm not talking about the need to have a full-blown laptop. Most tablets you can buy, whether it's an iPad or an Android device, have the ability to run software, apps or have direct browser access that enable all of those systems to integrate. All you need to do is put on a pair of headphones.

Recapping what I've touched on, there are a few obvious benefits from cloud computing.

Three of them are:

- Freedom: the ability to work from anywhere.

- Peace of mind: knowing that you don't have to worry about your IT.

- Increased profits: because all of your systems are integrated, you know that everything is running smoothly and efficiently.

Some of the common questions I get ...

What if I have many different cloud-based systems with the same data? Can they talk?

Majority of the time, yes, they can talk. Most cloud-based services and systems provided online have application integration (API). This is a secure connection enabling the different databases to talk and synchronise. If there isn't a connection to your cloud-based systems try getting a programmer to write a code that will make them talk. Most systems have a database, and it's just a matter of picking the fields so that they can match up and communicate.

An example would be a customer address. If you have an address in a customer database and want to synchronise it with your accounting system, it's just a matter of matching those fields and telling those systems how often you want those fields to synchronise.

What if I want to move one system to another system?

The majority of the time, you'll have a background database from which you can export all of that data. In the new system it's just a matter of matching up all of those fields to do an import. You may need someone to do some technical work, but moving between systems is obviously not a spontaneous decision and you'll have put much thought into it. If you haven't, it's time to sit down and really map out the pros and cons of changing systems, and figure out what the return of investment will be.

People usually move between systems to streamline the integration with other systems. This is because some systems have built-in connections to other cloud systems without the need of a programmer. Alternatively the people who made the software may no longer be in business, so it becomes necessary to find a system that's being currently supported and is flexible enough to suit your business needs.

I currently have a situation with a client who is using a system that does all of their accounts, quotes, and orders. However, the software company that created this system was bought by another company and they have decided to discontinue the original system. In this case, it has forced the business to either upgrade to the system created by the new company, or go through the process of finding another system that is suitable for their needs. Upgrading can cost tens – if not *hundreds* – of thousands of dollars, whereas cloud-based systems would be cheaper. Also, because of advances in technology, moving away from old software is simple, as it is easy to find cloud-based software to do what your previous system had been doing. It is dangerous to continue using unsupported software because if something happens (i.e. if there is a glitch, a software error or issue) you may not be able to have the problem fixed, which could leave the business without any type of system at all. Programmers would have to be involved, meaning possible downtime and incorrect sales reports, which in turn, will cause monetary issues for the business. Moving to a newer platform removes this risk, enables integration between the systems and streamlines the process going forward.

But the ultimate benefit to all of these changes is peace of mind. You don't want to have to worry about the on-site systems. You want to be able to put it completely out of your mind so that you can focus on developing your business.

Chapter Three
IT'S RAINING MONEY

W hat would you do if you had an extra 10 hours to work on your business? This is by far my favourite chapter because it shows you how you can make money by utilising the cloud.

The best thing about the cloud is that you know exactly what your costs are going to be. When you're all set up and you're running along, it's a monthly fee that only changes as you expand or shrink. It scales with your business. As you grow, so do your cloud services. It's simply a matter of upgrading them to the next level.

The majority of cloud services will have three basic packages: an entry level, a middle gold level and a platinum high level. The platinum high level usually gives you all you need to use the program, plus some additional options and add-ons. In the case of Xero, the add-ons are payroll employees, which allow you to scale the accounting package depending on how many employees you have on payroll.

Not all systems are the same, but this is how most have been structured. The best thing is that you know exactly what your monthly costs will be so you can work out how many sales you can cover with that monthly cost.. The time you're left with can be focused on working on sales.

You don't have to worry about setting up the system, integrating the system, or getting the system to work for you. It's already done.

You can then live the lifestyle you want because you're not worried about the IT side of things. You can work and make money from anywhere, with the flexibility of knowing all your systems are safe.

The biggest thing: you're saving time. Because you're not working on an actual system, or getting the system to work for you, it's good to go once it's set up. Once you have all the computers talking to each other, you can focus purely on communicating with your clients and customers, and getting quality products and services out efficiently.

The other great thing is that with all the systems connected and your focus on your business, you will be able to leverage your time and increase your productivity by 10 or 20 percent. This means you can focus on doing what you love. You're not mucking about with systems that don't work and you are able to work on what you want to sell, what you want to be focusing on.

There are three ways the cloud can make you money:

1) Increased reliability. Because your servers are not going down, your systems are not going down and uptime is set 99.99% of the time. You'll find that across a period of twelve months, you'll have saved on wages and productivity. It's a double-edged bonus.

2) There are no restrictions. Once upon a time, I had a business partner who was very technically fantastic. He would be able to adjust systems, get them to talk to each other, and build them from scratch himself. However, the

only issue with that was that he was a single source for technical work. If he wasn't around and something went wrong, there was no way to fix the issue. At the time, it was great to have him around because he was always there and always able to fix any issues, sometimes working late into the night. When our partnership dissolved, I needed to look for better ways of doing things within the business. In this case we had a ticket-logging system that was set up on a Unix server. I didn't know Unix and it was going to be a big commitment to learn it. I started investigating systems that could integrate QuickBooks, an on-premises accounting system. I found a cloud-based one that worked well. The downside was that QuickBooks was an on-premises system. The two systems synchronised and talked to each other, which was smooth and efficient, but I needed to log onto the server for it to happen. Admittedly I had remote access, but I needed specific software to run it and doing it on an iPad was cumbersome. I needed to look for smarter ways of doing things. Along came Xero, an online accounting service in the form of a cloud-based accounting system. Moving from QuickBooks to Xero was a reasonably large task because of all the accounts that needed to come across, and I really wanted to avoid having to run two systems if I ever needed to go back. Fortunately, my company hadn't been running that long, so there wasn't much that needed to come across.

Moving to Xero was a breath of fresh air. Xero has several other companies that have already created packages that connect into it. There I was, with a time and billing system, and a ticket-logging system that wouldn't synchronise easily with Xero. The system that we went to, and still currently use, is Zendesk, which unfortunately doesn't integrate directly into Xero. However, a system called Workflow Max sits in the middle. The integration between them is so easily set up that it's not a problem to have the three systems. Workflow

Max talks to Zendesk, Workflow Max talks to Xero. Because they talk directly to one another, I basically just had to say, "Yes, I have Xero." There were a few little things that I had to plug in, but after half an hour, it was pretty much all done. Once set up, the time and billing was now done through one system and automatically invoiced out to another, meaning nothing was missed and everything was streamlined. Moving to Xero also enabled me to integrate WooCommerce, an online shop interface in WordPress, which is what our website is made out of. Another program I've connected is eWAY, an online payment gateway system. This enables people to pay via credit card directly on the website. The bonus is that eWAY and WooCommerce connect directly into Xero and the systems talk to each other. All I need to do is focus on the products that are on the website and make sure that all the products are up to date. When we send out invoices via Xero, it lists the available payment options. This means the end user can click on Pay Now, which will automatically launch a form for their credit card details. Giving customers an easy method to pay their invoices with usually means they will pay them faster. This leads to fantastic cash flow.

3) Another great thing about Xero is that they've just introduced a follow-up email process for invoices that haven't been paid. As soon as the invoices go out, it triggers this system. If the invoice hasn't been paid in seven days, it will send clients a reminder notice. Once upon a time, Xero didn't have this function written into it, so I had to choose another system. In this case I used Vision 6, an online marketing email system. This also connects directly into Xero to assist in sending reminder emails. These are simple reminder emails that have a click option on them to pay the invoice right then and there.

Another high point is the fantastic online store integration, assisting with cash flow and following up debtors without me having to do anything. One of my distributors has online feeds that I can connect to directly. This means that instead of having to update products as the distributor does, the streamlined system will automatically update my website with the relevant details, including amount of stock available, location, product images and descriptions, using data inputted by the distributors. I have an online store with products that update themselves and all I need to do is make sure I manage the logistics side of things.

When moving to cloud-based systems, you'll find that you're left with a whole lot of hardware. You've got this server, which once had your accounting, email and file systems that you no longer use. Now you've put the file system onto Dropbox; you've put the email system onto Google Docs, Gmail, or Office 365 or a Hosted Exchange platform; and you've put your accounting system onto MYOB online or Xero or Saasu (just to mention a few).

Now you're left with this server. The thing is that people are still out there buying servers. This could be for many reasons. Maybe they're educating themselves to become IT administrators or maybe they want something cheap to play with. Some people still have old software packages they have to run in the background. So now you have this old hardware that you could possibly sell online on eBay or Gumtree, getting you some money back. You might not get much – maybe fifty or a hundred dollars – but in some cases hardware can get you a couple thousands of dollars because you've spent so much money on it.

Now you've moved all your systems to cloud-based, you've sold all your hardware, your systems are talking to each other, your staff now have less administration tasks, you have less administration tasks, and the focus is on the business. You now have 10 extra hours per week to work on growing your business. This means that you can focus on the marketing, investigate online marketing tools and sales, selling your product, upselling clients and getting feedback so you can improve your business. So many options.

Some of the common questions I've had are ...

What if I don't have a product to sell online?

The thing is that you might not have a physical product to sell online, but the tools and the marketing side of things really do revolve heavily around the internet. On your website you may not have anything to sell, but it would be great if you had something to give away for free to acquire a potential customer's email address.

That way you could market directly to them. You need something that will make visitors to your website say 'I want to read that and I'm happy to give you my name and email address to get that information.' What happens is they insert their name into the cloud-based marketing system. You get that in your database automatically, and then you can send a newsletter or dedicated marketing to that person.

In the case of an accountant, you may put out a free e-book or something light that people want to read. It may be simple advice on how to keep your debtors low, how to keep an eye out for the ideal client or even how to keep an eye out for a client that is potentially dangerous.

If your website offer concerns debtors, you'll be able to continuously send debtor-related emails to this person before gradually mentioning other services that you provide. This is direct marketing because that person has specifically given you their name or email address. Rather than sending a random email stating, 'Hello everybody, this is the information that I've got for the business', you can have an email that is specifically targeted and has that person's name in the subject line.

The other question I am commonly asked ...

What if I don't need to add remote aspects to my business? I have a physical business; I actually have to go there to work.

That's great. While cloud systems give you the option to be portable and mobile, they also give you peace of mind because their backups and updates happen automatically, and upgrades can happen quickly and easily. The accounting happens efficiently with tax table updates, and any other compliance updates that need to happen within systems will happen without you needing to intervene.

There are so many soft costs associated with upgrading hardware and software. The majority of the time we should be testing it first, so when an update does come out it should be run in a test environment.

The reality is that small and medium businesses don't have that sort of time or budget. However, these cloud providers have the time, budget and software systems to be able to integrate an update, and if something doesn't work, go back to the way it was before the change.

Mobility isn't the only benefit to using a cloud system; even if you work from an office you'll get peace of mind knowing your systems are working correctly.

What if I like being disconnected? I want to go away on holidays and not check email.

This comes down to personal preference. It's easy to switch off your device and to disconnect the email system on your iPad or iPhone. Put it into suspend mode, just don't look at it. I can't help you change your mindset, but I can give you the tools so that if you do want to work from anywhere, and at any time, you can.

The other thing with being disconnected is that sometimes emergencies crop up. What happens if you're on holidays and your accountant needs urgent information? The majority of the time, they'll have access to the accounting system because it's in the cloud and you've given them a log-in. Or something else might have happened on site, a failure or something that needs to be updated, changed, or reverted. Maybe a credit note needs to be allocated to somebody and you're the only person who has access to it. Whatever the situation, it would be extremely convenient to be able to manage it quickly and painlessly by having access to the cloud systems. Although you may want to be disconnected, it would be beneficial to have access to them just in case.

Chapter Four

THE SILVER LINING: SAVING MONEY WITH THE CLOUD

H ow would you feel if you could save thousands of dollars a month by moving to the cloud?

One of the good things about moving to the cloud is that you can not only make money, you can save money. Saving money affects your bottom line directly and reduces capital expenses. When moving to a cloud-based system, the cloud provider needs to make sure they maintain their hardware. They need to make sure that they have the latest servers, and hard drives are replaced as they fail. If one hard drive fails, it will not affect the full system. This means they can swap the failed hard drive out in a live environment without being noticed.

The point is that it is *their* problem; they look after it for you. As technology evolves, so do cloud providers. As new systems come out, they are doing the work to upgrade them and spending the hundreds of thousands – if not *millions* – of dollars upgrading their hardware infrastructure.

Solid-state hard drives have been out for some time and they provide a massive speed difference when accessing services. Cloud providers are moving more and more of their services to solid-state hard drives to provide faster systems for their clients. What would normally cost many hundreds of thousands of dollars for small and medium businesses now costs you sixty or ninety dollars a month.

In the case of a business with 10 employees, you may run Xero, an accounting system, for sixty dollars a month and then each of your employees has an email address, which is an extra 10 dollars a month. Instead of having a server costing you $20,000 to run those systems, you are now only paying $160 a month. You also have peace of mind knowing that the companies that are providing those services are updating their infrastructure constantly.

They also have redundant internet pipes and wireless options to keep connectivity constant. The difference with Cloud is that you don't need to go out and buy that $5,000 server. You also don't need to upgrade that server in three years or increase the memory. You've saved yourself thousands of dollars with a simple change to the business, so the bottom line is already impacted.

The other really good thing is that downtime is minimised. Your uptime is 99.99%; providing that with an on-site server is virtually impossible, and there are so many more things to take into consideration. Small and medium businesses just can't afford those failsafe options to be put into place.

Even a business turning over $100 million per year won't want to install a backup server because it is expensive. They're happy to take on the risk of systems going down. Moving forward, progressing to a cloud-based server and a cloud-based system for as many systems as possible will decrease their risk, decrease their potential downtime, and slightly increase their monthly costs.

However, for a business turning over $100 million with fifty employees and their on-site server goes down: that's fifty employees not doing work, being unproductive, who are still getting paid. Not only does the IT guy have to be paid to sort out the issue, all those employees are still paid to have a break. This could last half a day; restoring the server could be the only productive thing achieved. Working with those figures, let's say on average each of those employees is paid $50,000 a year. That equals around about twenty-six dollars per hour. Multiply that by fifty employees for about four hours work and you're looking at $5,500 in wages alone. Then you have to consider… are your clients happy? Are they getting the responses they want? Could you potentially lose clients by having half a day's outage? That client may be worth hundreds of thousands – if not *millions* – of dollars, so a simple outage could cost you an untold amount of money. Whereas a simple upgrade to a system over to a cloud-based system will ensure you won't have that downtime. You'll keep that client, your employees will keep working, and you'll still be making money – all without having to fork out the capital expense to keep it that way.

Another reason you save money and keep productivity going is that many of the systems can be used remotely. This means that some employees or you, as a business owner, may still want to work, even when you're sick. It may not be a massive illness, but you may not want to spread it around the office. Having too many employees off can obviously be very detrimental to a business. The ability to work from home without spreading disease can be very beneficial to a business, especially as a business owner.

When moving to the cloud, for the majority of the time, hardware is removed from the business. The hardware is on 24/7, which means it's drawing electricity twenty-four hours a day, seven days a week. When removing that server from the business, you've not only sold the server and made some money, but you're now saving on your electricity bill. When it comes to electricity, the majority of business expenses are incurred because computers and servers are running so often. By removing these, your soft costs decrease immensely. You're also saving money because of downtime. As I've mentioned before, having a cloud-based system guarantees 99.99% uptime, and means that you don't have employees not working and you don't have to pay the IT person as much to come in and fix issues. You will still have the odd issue with a desktop computer; however, those issues are minuscule in comparison to having a single-point-of-failure server running a business critical system, enabling the business to function.

Another reason you'll save money by upgrading to a cloud-based system is you won't have to hire a very experienced IT person. Obviously you want someone quick who knows what to do, but since you're not paying massive server costs you don't have massive downtimes you won't have to worry about employing the services of that really experienced ultra server technician to keep your business functioning. The cloud provider pays for that. They pay technicians $100–150k per annum to ensure their servers are running at their optimum. All you pay is $160 per month.

Some people really like having a server in the office. They like having control over that type of system, which is all well and good. However, if you really want to save money and protect the business, you need to move into the age that businesses are moving into. Most businesses now know nothing of servers; they only know cloud-based systems. If you want to stay ahead of the game, or at least with the game, it's a matter of changing your mindset to say, "I'm okay with not having a server in my environment because I'm saving myself thousands – if not tens of thousands – a year in lost wages, downtime and IT services."

What if the cloud goes down?

Sometimes the cloud goes down. There is no one hundred percent guaranteed uptime, which is why we talk about 99.99%. However, outages occur much less often than if you had an on-site server.

But how do you protect yourself from this? How do you protect your data that is in the cloud? Many systems offer an export option. In the example of Xero, this option enables you to download all of your accounting system into a neat little file or package. This way, you can always access and store your information. Other systems, like email systems, synchronise all your data, so that if you've got a mobile phone, tablet or computer with email access you'll be able to synchronise all three systems with the server. This means that if the cloud goes down, you still have all your emails on each of those devices. If you need to extract any information, you can do so from your emails.

Other systems – like Dropbox, which is used for sharing and storing files – are all synchronised. So although the cloud could possibly go down, you still have all of that information with the hard drive synchronised on your laptop or desktop.

To protect yourself, make sure you can download the information from the provider, or synchronise the data from the provider. The cloud is not infallible, so it's best to protect yourself as much as possible. Some people don't need this assurance, but others do. Either way, it's a good thing the cloud has options for you to synchronise and download the data you own.

Chapter Five

SOUNDWAVES
THROUGH THE CLOUD

Communication was one of the biggest reasons the cloud was invented. Its creators wanted to push as many communications into a cloud-based system as possible. When cloud-based systems are put into the cloud, they integrate better with each other.

Take cloud-based email systems; they integrate better with word processing systems, customer relationship managers, instant messengers and phone systems. Say someone sends you an email with a document attachment which you need to edit. Gmail provides Google Docs, which enables you to open that document directly from their document system. Office 365 has an online Microsoft Word option, so you can actually edit and view that document in Word through your browser. Again, no software is installed on your machine. It also doesn't matter whether you're on a Mac, PC or a tablet.

There's increased reliability, so you're looking at being able to access your data anywhere at any time, with almost one hundred percent uptime. The risk is removed, so you don't have to be concerned that it's not going to work for you. You can always change computers or work on a different system or platform if, for some reason, you have an issue on your actual computer.

Another reason for moving certain communications to cloud-based systems is they offer a stable budget. As mentioned before, for a fixed monthly fee you can get a certain number of email accounts, phones and instant messaging systems without worrying about upgrades, updates or management.

Some of the systems that you can get from cloud-based communications are email systems. Gmail and Google Apps are perfect examples, as they give you the opportunity to move your email to their systems. Office 365, which is a Microsoft system, moves Microsoft Exchange from on-premises to their cloud-based systems. Exchange Database, Microsoft's email system, is very powerful. They've developed it over many years and have reached a point where it does a whole lot more than simply send and receive emails. Other companies also provide hosted exchange systems. The difference is that hosted exchange can be in your country. In Australia, we have different providers that provide hosted exchange, so the email is actually kept and stored on servers that work in the same country, bound by Australian privacy laws and other laws around privacy and data.

You can also move your phone system to the cloud. Rather than having a big phone system in your home office, you can have a hosted PBX, which is a hosted phone system. Then all you need are actual physical phones in the office at your desks. These phones have the usual functionality that you would normally see in an office environment. They have call transfer, voicemail, and voicemail email, so if you get a voicemail it will be sent to your email address. This enables you to listen to it when you're out of office from your mobile device.

They also give you options for IVR, which is the system when you call in that says, for example, 'Press 1 for accounts, Press 2 for technical assistance, and Press 3 for sales'. The host provides it all and the majority of these services are quickly and easily changeable by the end user. There's a virtual office provider in Cheltenham that has moved all of the phones to cloud-based phone systems and a virtual PABX. They offer local numbers in all of the states in Australia, as well as the ability to log in and easily program and change the display names that pop up when numbers are called. This gives them the flexibility to go Australia wide, so that people in Sydney can call using a Sydney number and the office can answer the cost of the call.

The question comes down to reliability. Do you have a good enough internet connection to be able to run phones across the internet? Typically, these are called Voice Over IP or VOIP.

In a business, you can set up telephone systems in various ways. Typically one would set up a telephone system using ISDN lines, which are digital lines provided by a telecommunications provider. These can be costly but are highly reliable. Running your phones across a highly reliable internet service provides a similar, if not better, system than ISDN. The flexibility of not having to hold your own hardware is also a real bonus because, again, it's up to the provider to make sure they keep that up to date, keep it maintained, and replace it when technology becomes better. They also put in all the redundancy options. If one system goes down, other systems are in place. Fail over options like that are ideal for small businesses that cannot afford them on their own.

I queried some clients who run Voice Over IP PBX systems. I asked them what the sound quality is like, the downtime, and some of the issues they've faced. All of them said that the quality of service, the quality of phone conversations, and the uptime has been second to none. It is just like having a digital hard line service connected into their office, but with the added flexibility of guaranteed uptime and being able to change and add services as needed.

In the case of budgeting and keeping costs standard, it's ideal for this virtual office. They know that if they bring on a new client, and that new client needs a certain number of phone calls answered and a certain number of phone lines, they can pretty much guarantee how much it's going to cost per month.

From a communications point of view, social media platforms are another fantastic system. Facebook provides private groups. Setting up a company private group enables you to invite staff, subcontractors and contractors to communicate without the rest of the world seeing. A great thing about this is that if the employees are querying a certain product, or if they need to let everyone know the movements of certain employees (i.e. whether employees are off sick, etc.), they can do it simply through a Facebook group.

Other things include communication between employees and contractors, querying and upgrading processes, or querying an issue they might be having that can easily be fixed remotely, or by simply advising the staff on what to do.

A few systems out there, like Skype and Skype for Business, provide instant messaging services. These systems enable employees to communicate instantly via text. These sorts of systems can be integrated into the hosted PBX system so that you can tell by looking at an employee's status if they're on the phone, in a meeting, or unavailable. Simply being able to text an employee a quick message and get a quick response while they're on the phone can sometimes determine whether you win or lose a client.

Many businesses are now favouring telecommunication and teleconferencing over flying around Australia for meetings. Various businesses out there, including Cisco and Citrix, have systems that enable you to have online meetings between employees and external parties, as well as hold webinars for a multitude of invitations. These sorts of online systems can bring together a huge number of people without the need to fly all over the country and spend money. A client of mine who made $60,000 in one month by providing webinars to a multitude of people was down on the beach in Greece. This shows how effective a cloud-based computer system can be.

Obviously running a cloud-based phone system is heavily reliant on the internet. A good telecommunications company that provides a Voice-Over-IP-hosted PBX service will always look at the current infrastructure within the business to make sure that it will be correctly used. What you don't want is a business to jump too quickly into a hosted PBX system when their internet connection is inadequate. Most of the time, the internet connections will be separated, the computers and infrastructure will be on one system, and the voice and phone system will be on the other internet connection.

Some businesses are not in areas where hard line internet is available. In these cases, microwave links can be installed. These can be costly, so it's a matter of looking at the price difference between putting in a digital ISDN connection, or a high-speed, microwave wireless internet connection with a hosted PBX solution.

It's important to keep data within Australia on Australian-based servers. Part of Office 365's data centre is based in Sydney, but Google unfortunately doesn't offer an Australian-specific service. A hosted exchange system from an Australian company can be beneficial because that particular cloud provider will be bound by Australian privacy laws and governed by Australian laws.

This is very important, especially if the company is a finance company or a company offering certain financial information as they will need to provide information about the systems they use and where their systems are based.

CHAPTER SIX

BLUE SKY
ACCOUNTING

Accounting has evolved quite quickly from being software installed on your computer or server and run from within a business, to a cloud-based system. Xero has led the way in that initiative, moving very swiftly to a browser-based accounting system that provides integration into so many different systems, including payment gateways and various bank feeds.

The real benefit behind accounting in the cloud is again, mobility. It gives you the ability to log on from anywhere at any time and see your accounts at a snapshot. You can instantly find out whether they are up to date or mostly up to date, depending on who's doing the reconciliations and cost reduction. There's no need to pay massive, yearly fees based on how many employees are accessing the system. You now pay a simple monthly fee for unlimited employee access, which includes access for bookkeepers, employees doing invoicing, as well as accountants and financial advisers.

The cloud improves cash flow as automated systems follow up debtors, removing the need for actual employees to do the administrative tasks and forecasting. It gives you the ability to, at a snapshot, look at a business's performance, compare it to your goals, and set a budget which can automatically show you how you're tracking.

While Xero led the way in online accounting, MYOB quickly took steps to move their systems to the cloud as quickly as possible. Saasu is also another system that provides online accounting and integration. Meanwhile, QuickBooks is trailing behind but they're doing their best to keep up by providing a different sort of cloud solution. The user logs onto terminal servers to access their accounting systems, rather than use a web-based software-as-a-service type of system. But, in contrast to moving your accounting system to the cloud, the user doesn't have the same integration with various systems, meaning there isn't the same flow, ease of use or automation.

It can be slow and risky, having your accounting system run on an on-site server, as the necessary updates can increase costs. On a cloud-based accounting system, these costs are managed by the provider. They make sure that their servers and software are up to date and they automatically implement changes, tax tables and other requirements.

The four main benefits to moving to a cloud-based accounting system are: increased reliability, visibility across the board, integration, and the ability to have your bank feeds come directly into the accounting system for bank reconciliations. Having online payment systems integrated directly is very convenient. When invoices go out, all a client has to do is click on the link, pay by credit card and the money is in your account. Also ideal is the following up of debtors. This automated system reminds clients and customers that an invoice is due or overdue and prompts them each time for payment.

Other integrations may include CRM, so if the sales team enters a customer's information into the system, it will automatically update and add this data into the online accounting system. This, in turn, means the accounts team doesn't have to worry about finding this information or wondering where to send invoices and remittance, etc.

Cloud-based systems also reduce cost, as you will no longer have a massive server running your accounting system, chewing up IT resources, nor possible downtime. For fifty dollars a month, you will receive an online system with unlimited user access, almost a hundred percent uptime, automatic updates and easy control.

How do I protect my data? What if the cloud goes down?

Xero and other accounting systems give you the option to download your data, so you can do your own backup if you are worried about accessing it from the cloud.

Another queried issue is security. The biggest thing is the cloud provider; it's the cloud provider's job to ensure security is up to scratch. There's too much on the line for them not to maintain their security. It's a matter of making sure everything is up to date, all the firewalls, and that both hardware and software are in place. But it's up to you to make sure you have secure passwords and that you only grant access to people you trust. Only give them access to the systems or parts of the system that they require.

Chapter Seven
NESTING IN THE CLOUD

For most businesses, being able to access, edit, and send files anywhere and at any time is crucial. Cloud computing offers access to files in many different ways; some examples are Dropbox, Box, Trend Micro SafeSync, Google Drive and Microsoft OneDrive. Microsoft also has OneDrive for business. These online storage systems give you the ability to not only store your files, but also synchronise them with your local computer. This is a great way to automate your backup.

Another really good thing that cloud computing companies have started to integrate is the ability to edit files online. You don't need to have access to your computer, which makes this simple folder storage very, very powerful.

Applications like Google Apps and Office 365 enable you to edit documents and spreadsheets using a web browser instead of actually having to have the software installed on your machine. This is fantastic for businesses that use an Apple Mac at home and a PC in the office. The ability to be at home and edit files with Office 365, Microsoft Word and Excel through a web browser means there's very little difference in the editing process whether you're on a Mac or a PC. Being able to access those files from a mobile device or a different platform is also really convenient.

Cloud systems can also talk to each other. They may not

do so initially, but businesses work with each other to get the various systems talking. Microsoft Office 365 has OneDrive, a storage facility that is just like Google Docs. However, with Microsoft Office 365 you can edit stored documents directly. Most people were using Dropbox, which has a personal version with one terabyte of storage and a business version with virtually unlimited storage. At first, moving documents and spreadsheets from Dropbox to OneDrive so you could edit them was a cumbersome process, but the two companies got together and connected their systems. It's great to be able to edit your Microsoft Word and Excel files using Dropbox as your storage.

Other applications that you can install on your phone or tablet include options to have access directly to these file storage facilities. On an iPhone, for example, you can install the Microsoft Outlook app and connect your email system. You can also connect Dropbox, Google Drive and Microsoft OneDrive, so if you need to attach files to documents or save files from emails you can do so effortlessly because they're all integrated into the app.

Once you get these online storage facilities for files, you can use them for multiple things. Not only do they store your files that you can use every day, but if you're running WordPress as your website, you can connect a plug-in called Backup Buddy, which has an option to save the backup directly into Dropbox. This means that you can use your online storage facility to store your website backups. This makes it easy considering you don't have to purchase any additional online storage.

Some businesses may just need a backup for their business. Rather than getting the business version of Dropbox, they can install a network-attached storage device, which is like a cut-down server in their office. All workers will be sharing files through that one physical networked hard drive, then they just need to connect that network-attached storage device to Dropbox and tell it to backup and synchronise all of the files. This way they will not only have that on-site system, but will still be able to access their files through the company's Dropbox account in the event of a power or internet outage or employees needing to work at home.

By not using an online facility like this, you're risking the loss of files. You're also risking the loss of intellectual property, double handling and/or spending money on unnecessary backups (considering that most integrate into the single Dropbox account).

The three biggest benefits to having an online storage facility are:

- **Multiple uses.** You can use it not only for storing your files, but for various other online backups and synchronising services that require an online storage facility.

- **Peace of mind.** All of the files that you need on your computer are synchronised with Dropbox or Google Drive or OneDrive. This means that if something happens to your computer, or if you lose your hard drive or your files become corrupted, you can simply install the software on another computer and synchronise those files so you'll never be without them. Even if you need to jump on another computer to access the web interface of these systems, you'll still be able to access those files, and in some cases, edit them directly without having to install any software.

- **Flexibility.** Having systems that integrate with each other, and knowing that the cloud provider is constantly looking to integrate with other vendors and providers, means you're getting more and more functionality without having to lift a finger. The biggest one is the case of Dropbox connecting with Microsoft Office 365. Dropbox has done really well merging their system with others. It can integrate applications through various online and offline systems, enabling them to use their services for online backup, synchronising and general storage of databases and other facilities.

Issues that sometimes arise ...

- What if I don't have internet access? Do I need it to get to my files? The real benefit to synchronising your files with your local computer is that you can edit them offline and it will re-sync them when your computer finally has internet access again.

- What if the cloud service provider goes down? What if all of a sudden Dropbox loses my account or decides to close it? Well, since you have all of those files synchronised on your machines, you can just access them as you normally would, setup your account again and resynchronise them.

- What if a file becomes corrupt or I save over it with an older version? The great thing about these cloud providers is that they do version control. Version control means that most of the time when you hit save or overwrite a file, it will remember the previous versions. Say, for example, you've worked on a file but accidentally saved an email attachment over the top of it. You can go back and revert to the originally saved version. This also means that if you accidentally corrupt a file, you can go back to a previous working version. You get the added flexibility of version control that you'd normally get from an on-site server, but without the massive cost.

An obstacle with these cloud-based file systems is that it's difficult to have two files open at the same time. Say, for example, a business of 10 employees all have Dropbox installed. They all have the same file synchronised and Joe and Bill decide to open it at the same time. They will edit their own edits and then save the file.

One of these will be a conflicting file and won't make it into the file system as it normally would. In these cases, one would normally install a network-attached storage device, which is like a cut-down server, into the environment. All employees at that location would have access to it, allowing them to open, edit and save files as one normally would from a networked drive. The difference is that Dropbox can then be used to synchronise those files into the cloud. Network-attached storage devices prevent two people from simultaneously editing a file. If Joe opens a file and then Bill tries to open that same file, Bill will simply get a notification informing him that he is unable to edit that file until Joe is finished.

In this regard, businesses with multiple locations can find themselves in a little bit of strife. This is where online services such as Office 365 and Google Apps come in handy because they allow users to both access and edit the same file at the same time – just not the same specific area at the same time. If you're in Excel and you're editing cell A1 and someone opens that same Excel file, the only thing they *can't* edit is A1; however, they can edit the rest of the spreadsheet as needed.

Google Docs offers a very similar option, whereby two people can edit the same document or spreadsheet at the same time. This is very handy for collaboration and for businesses that have multiple locations and really want to keep as much as they can in the cloud without installing any software on any computers. It's also useful for running multiple operating systems (across Mac, PC, various Windows machines, etc.) and for accessing the same software online simultaneously.

CHAPTER EIGHT
LIGHT WORK

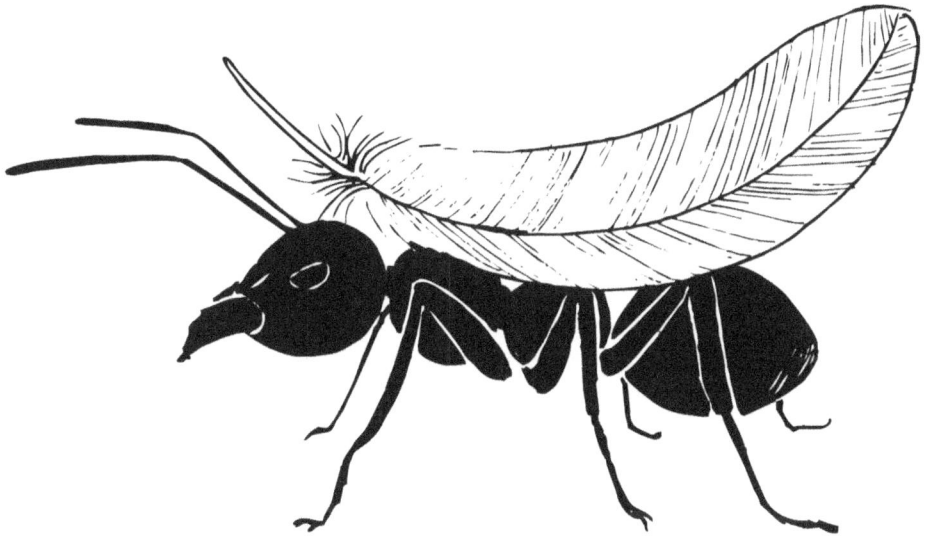

I 'm going to let you in on a little secret: moving to cloud computing reduces your IT support costs. It almost removes the need for IT support, which isn't great for the IT support guys out there, but it's fantastic for businesses who want to save money.

The best thing about moving to cloud services is that the cloud provider usually looks after everything. They look after all the upgrades, all the maintenance, all the hardware checks, all the ongoing support, and any integrations and assistance you need done.

They offer, most of the time, unlimited support based around their package. Multiple people out there are submitting information on forums about how to do various things yourself with these cloud systems. You get quick responses, because they have the budget to make sure you get the support.

What does that mean for small and medium businesses? It means the money can be better spent. You can focus on spending money on marketing and sales and growth within your business, rather than on capital expenses and other IT expenses that would normally be quite substantial for a business wanting to move into business-related systems.

Many people don't realise that the software that makes these cloud-based systems work have various failsafe systems in place which make it easy to roll back to a system that worked before the update. Say, for example, there's an update that's going to be released for your online accounting system. The online accounting system will trigger an update that will go out to a certain group of people using them as test subjects. If everything works okay then it's released to the rest of the organisation. If the update causes issues on that system, it's automatically rolled back to the developers, so the developers can continue working on it to ensure it works going forward.

Although small and medium businesses should be testing updates before rolling them out, this is not always cost effective and can cause time issues. Fortunately, cloud providers do all this for you, meaning you can enjoy a fantastic experience for a very small cost.

What does that mean for business in terms of support costs? Well, you won't need a server or experienced technicians to manage your server. Most issues are computer-related and usually occur because something online has been clicked that shouldn't have been clicked, causing a virus or some sort of issue. Otherwise they're caused by corrupted updates from Microsoft. These may cause grief, but they cost less to fix than a server issue that has stopped your entire business from functioning.

Most of the time an employee that's using one computer can jump on another or a laptop and do at least some work from an iPad or another device. They can at least limp along while their original computer is getting repaired.

The other great thing is that you don't need the technician to actually come out onsite. Sometimes there's a callout fee involved when a tech support is required to come out onsite. However, most of the time we can send out a courier to pick up that laptop or desktop machine, send it back to the IT service depot, have it fixed and returned to you for substantially less than you'd pay for a callout fee.

Most providers also have a remote desktop or a remote control session that they can use to remote in and fix whatever issue you're having. The good thing about this is that you can actually see what they're doing and keep track of how long they're taking to get it done.

Most IT providers work on a pre-paid hourly basis for any ad hoc work that needs to be done. This is because it's quite cumbersome to have to invoice out in fifteen-minute blocks and invoice fifteen minutes here and there. If you simply buy a pack of five or 10 or 20 hours (depending on how much work you'd normally need on each of the desktops) it simply just chews through that.

Although this seems like a large amount to have to pay initially, if you look at the administrative costs associated with processing fifteen-minute invoices you'll see it's so much easier to accumulating various services over a few months and then at the end spending five minutes glancing over all the work that's been done and working out what needs to be adjusted or investigated a little bit more. After a couple of months you should be able to work out how much work is needed. You'll be able to tell if you need the five, 10, or 20 hours going forward.

Some IT providers also use retainers, which usually cover unlimited support. In our case, our unlimited support arrangements depend on how many computers the client has. The cost is per computer, but this also includes an antivirus and any necessary upgrades, which assists in keeping the budget stable. The majority of time, the idea for businesses is to keep the budget stable. The last thing you need is a massive unwanted invoice.

In short ...

The cloud providers do most of the hard work for you. They assist in providing the integration for other systems. They provide all maintenance and support; you simply need to pay the monthly service cost. The need for highly skilled server technicians is lessened and the business would only require a skilled desktop support team going forward.

Another positive aspect to having a retainer or an ongoing relationship with an IT business is education. Quick phone calls, video requests or webinars from the IT provider are often included in your monthly retainer. This is ideal as it keeps everyone in your business educated about what they should and shouldn't click on and how to use certain applications. It's also a good opportunity for them to have any other general questions answered. This keeps the education going and the touchpoints with that specific IT provider.

Because PCs and Macs operate differently, there isn't always going to be one provider that will suit you. I would suggest reviewing the contract with your provider every twelve or twenty-four months and assessing the relationship. Do you still have a positive working relationship? Are they giving you great value for money? Are they going above and beyond? Do they send you how-to videos? On the various systems that they provide, are they offering other suggestions? Are they telling you about new technologies out there or integrations with other systems that you already use? You need to ask these sorts of questions to ensure you're getting the best deal and growing your business with the best technologies.

Different cloud services are very similar to telecommunications. Telecommunications companies may change their plans without necessarily telling you, meaning you may be getting the same deal you're getting now at a higher cost. IT companies, telecommunications companies and cloud service providers should be treated no differently from each other. Always check to ensure that you're on the correct rate as there are always changes within systems to keep them at the best rate possible.

What if you don't like your IT provider? What if you'd like to move away from them but you're worried that they have so much access? It's always good to engage them face to face and not through email. Text messages and phone are probably best. That way you can bring them on, they can look through your systems and make sure that they are capable of understanding and assisting in the various systems that you're running. Sometimes IT providers also provide the cloud solutions, so it's a matter of figuring out what you've got access to, any portals that you need access to, any passwords you need access to so that you can change those. Most of the time IT providers get money by providing the cloud service so as long as the cloud service is working really well, having a separate IT provider to do your IT support will keep costs low.

Not every business is the same and sometimes a business will look to change IT providers *and* cloud service providers, depending on what services they have. In these cases it's best to look quietly at other IT and cloud service providers, and work with them to migrate across to their service.

Most of the time, IT providers will use templates to perform a migration, especially when another IT or cloud service provider is already providing these sorts of services.

One thing you need to be safe and secure in is that it's your intellectual property. It's your data and you should have access to it whenever you want without repercussion. It's in your current IT provider's best interests to ensure that a termination of services is done as swiftly and amicably as possible. It's never nice losing business. However, if you're not providing the right service, be prepared to change and improve. If that means you lose some customers, it will at least force you to provide a better service.

CHAPTER NINE

THE CLOUD DOESN'T CARE

W hat would you do if your computer or laptop was stolen? What would you do if your office computer or hard drive just died?

Would you be able to get a new computer and resume working straight away? Would you be concerned that you lost some data, or would you simply connect to your cloud-based systems and re-download everything?

The issues I've come across regarding laptops and desktops are hard drive failure and hardware failure. Sometimes we do have the misfortune of having something stolen, but the majority of cases that I've come across involve physical damage to the machine. I had one client run over her laptop with her car. I had another client whose server room was flooded and they pretty much lost everything from water damage. I've had Mac hard drives die, PC hard drives die – all sorts of issues with computers.

But it's not just computers; we also have our phones and our smartphones. We have our contacts, our mail and our photos; so many precious memories. The amount of cherished data we have on our devices is massive. Losing that data can have detrimental effects to our mental wellbeing and our business.

The introduction of cloud-based systems to mobile devices

has really been beneficial when it comes to backing up our data and ensuring we don't lose precious memories or information, or the data we need to run our business.

So the cloud doesn't care. The cloud doesn't care what sort of computer or mobile device you have. It doesn't care whether you're on a Mac or a PC; whether you're running Linux, Windows or Mac OSX; whether you use an iPhone or an Android mobile device. Cloud-based applications are compatible with all of these, enabling you to back up and store information to all of these devices.

We can run our accounting system from our Mac, our PC, or run it from a mobile device or tablet. It doesn't matter what device we have because the developers of this software – the creators of this cloud-based system – made it so that you can just run it through a web browser. They create the apps and the interfaces that enable you to have massive flexibility.

I have some clients who get their employees to use their personal laptops rather than buying them computers. As an employer you can say to your employee, "You know what? Bring that personal computer in and we'll run all of our systems on it just by using a web browser. All we'll have to do is set you up with all our PINs." It means that employers don't have to worry about their capital, but can know their data and information is protected because of the security levels on these cloud-based systems.

Event tracking and log-ins ensure you can track where and what happens on your cloud-based system. You might have an accounting system where someone sets up an input and sends out an invoice. All of the details are attributed to that particular employee, so you'll know who sent it, who created it and all other relevant details. This means you can go back to an employee and say, "Look, you've done this incorrectly. Let's talk about training you in how to do it properly."

The main systems are accounting, files and email. By running an email system through Google Apps, Office 365 or hosted exchange, you can use a web interface to simply log into that email system. All of those systems can connect into Dropbox or your file system where your files can be added or edited through a web interface. Google Apps also has this option, allowing you to seamlessly edit your files through a web interface without having to download anything. You have your files, your email or accounting system and you're running in a cloud-based web interface. That's all you'll need, regardless of what accounting system you're running. You won't need to download any software and it runs seamlessly.

To run those three main systems mentioned earlier, your employee simply needs a web browser. Web browsers are Google Chrome, Internet Explorer, Microsoft Edge, Firefox, Safari – there's a multitude of them. And developers are always finding ways to make them faster and better.

These systems also make it easier to use a tablet or phone device. This means you can access these systems directly from your mobile devices or tablets while you're out and about. For example, you might need to access your accounting system from your iPad. You would simply open up Safari, type in the address, log in and you'll have full access to that accounting system. You'll also have access to your online banking systems as well as your online mail, your files directly from Dropbox, Office 365 and Google Apps. All of these systems were created to make your life easier and more mobile. Now you can take that trip with the family or take your device overseas and work from virtually anywhere in the world with an internet connection.

You might find you've been working with computers, Windows, PC and are looking to make a change. You want to move from PC and Windows over to Mac, Macintosh, Apple Mac, OSX. Cloud-based systems enable you to do this really easily because there aren't many files stored on your actual Windows machine's hard drive. You may have documents, spreadsheets, maybe photos; however if you've synchronised all of that with Dropbox you can connect to Dropbox on your new machine and re-download those files. This is very easy to set up and you'll have everything working again very quickly.

However, if you were in a hurry and wanted to get all your systems up and running as quickly as possible – say your hard drive had died on your Windows laptop and you wanted to replace it with a new Mac computer – you'd simply use the web interface to manage your accounting system. This way you could do things like send invoices, send remittance and bank reconciliations all without needing to download any software.

Three of the biggest reasons why the cloud doesn't care:

1. Use any device, any operating system. The cloud doesn't care what you're using, as long as you have an internet connection and a browser.

2. Reduce setup time. The cloud is ready to go straight away. You get your new computer, jump online, open your browser and you're good to go. Log in and continue where you left off.

3. The cloud-based options are scalable. If you're a start-up business and you need to run as lean as possible, show some forward thinking, look at the

bigger picture and opt for systems that will work for you in the long run. However, you might not be able to spend that sort of money so you need to start smaller. Many of these systems, whether they're Dropbox or an accounting system, have an industry-level product. They offer products for free or for a very small monthly fee. These particular systems can grow with you. So as you bring on more clients and staff, as you process more transactions, as you have more emails coming through, etc., you can upgrade incrementally without needing to jump from nothing to something quite large. You can slowly build it up. You can usually choose from three or four different plans, starting with the basic free or small monthly fee before upgrading to a model that gives you more options for a larger fee.

Earlier I touched on mobile device backups. Apple's iCloud service is fantastic. It backs up your entire phone and can synchronise all of your photos, including any new photos that you take, with Apple's online storage. Dropbox also has an app that can go onto the phone so that you can synchronise your photos. If you don't want to synchronise them with Apple and iPad you can sync them directly with Dropbox. This means that your phone will upload any new photos you take directly to your Dropbox account as soon as it detects a wireless network.

The iCloud's great because it can back up your phone in its entirety. This means your text messages, applications and any data related to those applications will be safely backed up for you. You can elect to turn on the photo sync or leave it off and connect that to Dropbox. There are a couple of different options there.

Android also has a backup option. Connecting your Android device to a Dropbox account is ideal. You can also back up your phone to Google Drive – a service with various options. Again, you can do this on the fly, without having to physically plug your device into the computer. This is where many devices are headed. Many people almost exclusively use tablets and their phone. They don't have a computer at home, so it's essential for them to have a cloud-based backup. I've had too many clients lose their phone, have their phone stolen, or accidentally wash them and have to start from scratch. They lose years and years of photos and there is no easy way to retrieve them, so please back up your mobile devices. Spend a bit of money. If you need fifty gigabytes or one terabyte worth of cloud storage it may cost you a hundred dollars a year, but the alternative is losing every single photo on that particular device. So you can either spend a hundred dollars or risk losing those priceless photos. You can't put a monetary value on that.

As I said before the systems we're running are scalable, meaning you can start with the free version and move up to a paid version. Many of these systems can also move back. As you increase staff you may find you have to pay for extra services, but then as you reduce staff you can reduce your additional services as well. So if those particular employees aren't generating billable hours or enough income you can remove the parts of the system that they don't use to reduce your monthly cloud payments.

As a takeaway from this chapter I would request that you look at how you're running your business and what systems you have that can be moved to cloud-based systems.

If you have software installed on your computer that's running a certain system – like, for example, an accounting system – I strongly recommend looking at a cloud-based accounting system. If you replaced your computer with a brand new one, how quickly could you get it up and running? Would you have to get somebody to forensically recover your old computer's data? Or would you simply restore the applications you need to install, synchronise the files you need to synchronise and be up and running within a matter of hours or even seconds, depending on what systems you have moved to.

Say, for example, you're running a customer relationship manager. That piece of software sits directly on your Mac and contains the contact details of all of your clients, all of your contacts. You may want to move to a cloud-based system, but are wondering how much it will cost to migrate all your data from your Mac into that particular system. Most systems have an exporting function, meaning you can export all of those details into a file and then import the data into a cloud-based customer relationship system. Many of my clients move between different cloud-based systems in this way. All they're required to do is export their data and import it into the new cloud-based system. They often move to a cloud-based system to share data with their employees. Sometimes they need their employees to assist in updating their contacts on the fly or they may just want to synchronise those contacts with their accounting system.

It's recommended you remove as much redundant data as possible and have systems talk to each other. This way, if your bookkeeper updates a certain client's contact details the system will notify your CRM. If you have an app on your phone that connects to your CRM and you go into it to call that client, you'll know those details have been updated. Or say you're having a meeting with a client and they tell you they're moving offices, you can put their new number directly into your CRM, which will then sync with your accounting system so that the number will be correct for all bookkeeping purposes.

While the cloud doesn't care what sort of computer you have or what sort of operating system you're running, you should definitely care about what you're doing with your data. How are you backing up your data and how do you treat your devices? Make sure you have a backup in place and that you're synchronising your files and doing tests every now and again. Don't just assume that everything is working perfectly because it's in the cloud. Always put a file somewhere to check that it is syncing across all your devices. Sometimes things happen, but if you are vigilant, stay on top of things and do continuous checks to ensure everything's running okay, you're in a far better position than if you had everything stored on your computer.

CHAPTER TEN

AVOIDING THE STORM

What if you never had to see your IT guy again? One of the many benefits of the cloud is ongoing support. This allows you to utilise systems by yourself without the need of a specialist IT person. Having your own in-house server and applications means you have to support them by employing someone or engaging an IT company.

Some of the downfalls of running your own server include maintenance and having to make sure that the updates are done. Unfortunately for small businesses, updates can't be done in a test environment; they have to be done in a live environment, which can cause many unknown issues and cause different software to stop working. Sometimes updates cause more grief than there're worth. Recently Microsoft released two updates that have caused issues with Microsoft Outlook. If businesses had a test environment to roll out in, they would be able to see these problems before going into the live environment. Unfortunately small businesses don't have these luxuries. They have to roll out updates into a live environment, which can cause downtime.

Having your own server means these risks are higher. In-house servers for small businesses are a must. There is no other option. This is usually because the file system can't be accessed through a cloud-based system, as too many people in the office are trying to access the same file at the same time.

Various applications, like Google Apps and Microsoft Office 365, are starting to overcome these issues, allowing multiple people in and outside of your organisation to edit the same file online if they have permission to do so. However, some of these may be quite large CAD drawing files or they might be graphic design or movie files that are being edited at the same time. These are being saved and uploaded to the server simultaneously. Doing this across an internet connection is not advised and would cause more grief than it's worth. However, accessing these files through an internal network and internal server will improve on speed and efficiency.

Additionally, the cloud also assists with these setups. Every night a cloud-based backup system will be performed on the server and in turn, backup all those files.

Having your own databases is another risk to businesses. An example of this is the Microsoft email service Exchange, which runs off a reasonably sophisticated database; the more databases that can be removed from your server and pushed out to cloud-based systems, the better. In the event that an on-site server crashes, it's much easier to download the files to a server file system than to reinstall an application and reattach a database. In some cases, it's almost impossible without a huge expense and effort from IT professionals.

I've found that the less money small businesses spend, the happier they are. Giving them options that remove the risk of massive expense when something goes wrong is far better for the business. Not having a server can bring large savings to a small business. Many people don't realise that they pay for a whole lot of peripheral expenses when they have a server. These can include the maintenance for an IT company or an IT professional. On the hardware front, not having to buy a server means no capital expenses, and this means you can focus on operating expenses within the business using cloud-based systems.

You also won't use as much electricity. When running a server, you don't come in over the weekend to turn a server on in the morning and then turn it off when you leave. Most of the time it's on twenty-four hours a day, seven days a week, which puts a strain on the hardware and uses excessive amounts of electricity. The moving parts generate a lot of heat and require air conditioning units to keep servers cool. This prevents them from melting the CPU and hard drives, which can cause loss of data. On top of having to power the actual server, you have to power an air conditioning unit. Again this can be quite costly.

The more outgoing expenses we can save on a small business, the better. By moving simple services that one might use on a large server to a cloud-based system, you'll save massive amounts of money.

Many IT people out there still don't trust the cloud. As fantastic as the cloud is – I know various app developers put great effort into their security – there are also processes that you can put in place to synchronise and remove redundant data. This way if anything happens to your cloud provider you will still have all your data.

For security and peace of mind, there are various synchronising and encryption options to protect you from having any data stolen on cloud-based systems. The IT people who are uncomfortable with the cloud need to realise the potential benefits it has for small businesses.

Once upon a time, a business might have spent $10,000–$20,000 on a new server, which could last anywhere from three to six years. This depended on how much they pushed that server, where it was physically located and other variables. Business owners also had to cross their fingers that the hardware would last that long. A $20,000 server is profitable for an IT professional and it gives them ongoing business. Moving forward, they have ongoing maintenance contracts that are common with servers of that size and they're guaranteed work because Microsoft updates will always be needed. There is also the potential of failure, as in the realm of IT nothing ever goes to plan. For a small business, forking out that much money is difficult. Cloud-based systems, however, are usually subscription based. This means you only use what you need. If you only have 10 employees, you only pay for 10 licences. If you downsize to five employees you can reduce your licence level. You know exactly how much it's costing you per employee to run your systems.

The IT guys who want to keep themselves in business will usually push to sell a server. Large servers are usually unnecessary for a business. Microsoft is pushing more and more of their systems to the cloud to help small businesses become mobile. This way, small businesses aren't locked to a desk, location or internet service provider. It doesn't matter what internet service you use; you'll still have access to your systems.

Cloud-based systems are becoming more and more advanced and can now talk to each other. Microsoft now works with Dropbox, Dropbox also works with Google, Google has their own similar system called Google Drive, Microsoft has their own system called OneDrive, and Apple has their iCloud and various other systems. Dropbox also integrates with your Apple products, giving you synchronisation options from your PC, Mac, iPhone, tablet or Android devices. Software has evolved substantially so you can connect up multiple systems to access your data.

Another fantastic thing about cloud-based systems is people are developing systems that enable connection to other systems. An example is Xero, the accounting system. Xero is very flexible and has various application integrations, or APIs, which another company, Zapier, has implemented to give businesses and individuals power and flexibility with their data.

Having a single point of connection with all of your systems is invaluable for small businesses. For example, when I enter a new customer into Xero Zapier automatically detects them thanks to various systems I have set up. This then triggers a synchronisation of that data from Xero into an email system. In this case it's Vision 6, but you may be more familiar with MailChimp or Aweber. It puts the information into that system, which then detects a new person and sends them a welcome email. This welcome email may have information tailored for that specific client, including passwords and log-in information.

I have set up another trigger, which notifies me that a new person is in Xero. I then set up their username and password in our ticket logging system Zendesk. This is another cloud-based system, which the client can log into to log tickets and see their support tickets. They can also look at their knowledge base and search any questions they have before making a support call to see if they can fix the problem themselves.

Zapier has many trigger options and actions, allowing you to more or less automate your entire business based on the sole action of entering a new client into Xero. This then triggers various other systems to run and complete their tasks. This cuts down on admin, as you won't need someone to log into each of these systems to manually enter the information for the new client. It also benefits the client relationship as the client will receive a lot of useful log-in information via email. You might even have some induction videos for them that you could put in an email or on YouTube.

Zapier gives you massive amounts of power and the ability for small businesses to look much larger than they actually are. When the clients are happier, they'll want to spend more money with you, and they'll be more likely to take your advice and listen to what you have to say because you have engaged them so well.

With cloud-based systems, the cloud providers keep everything up to date. Whereas if you have your own server, you have to engage an IT company or IT person to assist in keeping systems up to date. Because it is their job to ensure you have the best possible experience, cloud providers do their updates on test-based systems to ensure they rarely go wrong.

There's also software out there to help programmers streamline the rollout of software updates. They might rollout an update through their software. The software will roll it out to a shortlist of clients and it will then detect whether they are having issues or things are running smoothly.

If those clients do have issues, the software will automatically roll back to the previous version. No data is lost, no downtime occurs, and the applications continue to provide the service that is being provided. This is while the update is being prepared and a new version is being updated and rolled out. These systems enable small and medium-sized businesses to harness these test environments, without needing to fork out money for duplicate systems.

Cloud-based systems will also always grow and develop. Most people don't realise that with software or accounting systems as big as Xero any small change will have a domino effect and will be replicated throughout the entire system. One must take this into consideration; a change as small as moving from two decimal places in an invoice to four decimal places can affect the entire system. Having a report display correctly on the screen with four decimal places instead of two is a large job that obviously requires recoding many reports. A lot of testing needs to be done to ensure that those reports don't go off screen. They can look quite presentable in a printed report and on screen as well. This system domino effect means a lot of testing is required before an update can be released. But these cloud providers are always developing, always looking to better their system, whether it's to increase their search options, add additional functionality, or set up integrations with other systems to enable you a better customer experience.

They will always be updating their systems.

A question I often get is "What if I like my IT guy?" That's all very well and good; you can keep your IT guy to ensure everything runs as smoothly as possible. There are going to be instances where Outlook doesn't open correctly on your PC or a virus gets into one of your computers or laptops, or one of your sales reps loses, drops or destroys a laptop, so a new one will need to be set up. You will always have those little expenses, but the main thing is you won't be solely reliant on that IT person. You have a good relationship with them, they assist you with your IT needs, but if they can't fulfil your needs or they're not around you'll be able to figure out some of these issues yourself.

The bigger systems that your business relies on – where downtime means loss of revenue, sales and profit – are looked after by multimillion-dollar companies whose sole business is to make sure they're continually running.

What if you don't want to use systems outside of Australia? What if you want to keep all your systems inside Australia so Australian privacy laws govern them? You want to know that if something goes wrong you can tackle it legally in Australia. Most of these systems have options in Australia. All the data is kept in Australia so you're able to access a similar system that is on shore. In most cases, this is more expensive; however, it's not going to break the bank and will still give you many benefits. In many of these instances, these systems give you an option to synchronise your data with your computer or download it so that you have a local copy of your information if you so wish.

I remember my first meeting with a new client I brought on board. I'd never seen so much relief and excitement in someone's eyes. His previous IT guy's heart was in the right place; he gave my client free systems and software that tackled spam filtering, but these ended up causing more grief in the end, as he was the only one who could assist with the issues. My client was very intelligent and extremely knowledgeable about general IT. But even he couldn't understand why things weren't working. Why they were getting viruses? Why was the email system going down? And why weren't the computers talking to the servers correctly? A mess of issue after issue was caused because of the IT guy's free systems. These free systems required knowledge of configuration and a lot of upkeep in the form of updates. It demonstrated why free solutions from the internet aren't always right.

I ended up moving my client from a server to a cloud-based email system (an Office 365 package) because they wanted to connect SharePoint with Dropbox as well as harness the power of the email system and Microsoft Office packages. I got rid of their server – in this case, they had some files that needed to be kept locally, but these were backed up after hours in a network-attached storage device. This streamlined their business astronomically. They went from having to regularly pay someone and endure the frustration of systems going down all the time to barely needing to call us. The few times he did call were for advice –"We're looking to bring on a new system, or integrate our current system with something else. Can you assist?" or "We're looking to streamline our business. We want to start taking online payments or start doing something else." He moved from the frustration of "Why isn't this working?" to "This system works so well!" to "What can we do next? How can we improve the business? How can we improve cash flow?"

The whole conversation changes when clients move to a cloud-based system. I'm no longer called to hear about clients' issues, frustrations and anger; I'm called to share in their happiness and inquisitive nature. It's a great feeling to be able to move IT out of that realm of frustration and show clients the excitement it can actually bring to their business.

CHAPTER ELEVEN
MIGRATION

Iknow what you're thinking, but it's easier than you think. The majority of small business applications can be run from cloud systems. These include email systems; customer relationship managers; content management systems, which are websites and managing websites for accounting systems; email broadcast systems; other marketing systems; and point of sale systems. Most of these have cloud-based options.

Cost is a small business's biggest concern. Small business owners often wonder how much it is going to cost them to migrate to a cloud-based system and how this cost compares to running their own server or upgrading to a newer one.

It's a good question to have, as it's important to look at how figures will affect a business's bottom line. For example, if you had to buy a new server and the server cost $5,000 you have to factor in the actual installation, which might cost an additional $3,000. This means you're up for about $8,000 worth of capital for the business.

A cloud migration, on the other hand, might only cost you $2,500 – or even less, depending on the size of the business. This is quite a large saving. You then have ongoing monthly fees, which you can claim back as an operating expense for the business. Depending on which system you go with, this can be a whole lot cheaper than $5,000 a year. It could even be cheaper than $5,000 for *three years*.

The other thing we need to look at is ongoing maintenance. Once you move to a cloud-based system and pay the monthly fees, you won't need to pay any extra for maintenance. The company providing the cloud service maintains the system and ensures that everything is upgraded and up to date. For accounting systems, they make sure that the latest tax tables are loaded from the Australian Taxation Office.

All of those components that you'd normally have to pay extra for are all covered in that monthly fee. While a capital expense for a new server install might be $8,000, you'd also have to take into consideration the maintenance that's needed to keep the server running, just like you would when maintaining a new car. Most of the time this is a monthly fee, so in the long run opting not to go with the cloud will cost you more money and give you less benefits.

By going cloud you'd save yourself the capital expense and the monthly maintenance required to keep a server running smoothly. You can then redirect this money to your local workstations, which will ultimately require less maintenance because they're now protected by proactive services, like monitoring systems that check and look after hardware and ensure the antivirus system is up to date. This means you'll have less downtime because hardware or software issues are being picked up before employees even notice them.

Also because projects can be very similar we're able to use knowledge gained from past projects to streamline future projects so that they can be tackled more efficiently. There's also software that has been developed by various vendors to assist in the migration process. This makes the migration from one email system to another much more efficient than it used to be and it means little to no down time for the majority of the employees within the business.

We're basically looking at a cookie-cutter approach to cloud migration. Whether it's a file system or an email system, the software and steps make this happen as seamlessly as possible.

What if something goes wrong during the migration process?

Things often do go wrong; it's never one hundred percent smooth sailing. In saying that, though, it's easy to create a workaround for most issues. For example, if someone's mailbox doesn't come across to the new mail system easily, you can usually bring it over manually. This might mean a delay of up to an hour, depending on the size of the user's mailbox.

I'm a small business with a small budget. How do I move from an inexpensive on-premises system to a cloud-based system?

Most of the time, you won't need to move your entire server all at once. There are options to do it step by step – i.e. doing the email, file and accounting systems separately. I recommend this step-by-step approach because sometimes the IT people need to work with the accounting staff to move the accounting system to an on-site system. This can take a bit of time, so it's best to approach it in stages.

Stage One may be the accounting system, Stage Two may be the email system and then Stage Three would be the file system and the retiring of the server. Doing it in stages like this means your budget can be spread over a certain period, which can be ideal if you're operating on a smaller budget.

One of my clients desperately needed to move from an on-premises server to a cloud-based system. They managed to prolong the life of their server to about nine years but at a certain point it starts to fall apart. Because of this, it becomes very risky to keep that server running, so moving to a cloud-based system as soon as possible was imperative for this particular client. We did this in stages. First we looked at a suitable email system for them. Pulling the email system off the server was the most crucial part of the project, so that was done over a weekend and they were able to work using the new email system as of Monday morning, working through some of the changes with each of the employees. Most of the new systems worked the same way as their predecessors – except they were now on a cloud-based system rather than on premises. Most of the employees didn't notice any difference at all.

The next phase for that client happened six months later. We looked at moving their file system to the cloud to enable employees to share files and collaborate with each other. Once again this happened over a weekend. This was great for this particular client because it worked within their budget. Six months later we looked at moving their accounting system to Xero.

It didn't happen all at once. The most risky parts of the server were removed and migrated to a cloud-based system first and the less risky components were done at a later stage. In an email system a database is quite complicated. But having this in a cloud-based system that is maintained and managed by a cloud-based provider means you aren't reliant on the on-server database. Restoring a database after a server fails is a large job, but files can be restored with just a backup hard drive. Restoring failed hard drives would be easier and faster for this particular client than a server crash, where files would be irrecoverable. Fortunately, because we approached it in stages it was not imperative that we move that straight away.

The most important thing was the email system with the email database. We moved that to a cloud-based system and would look at safer systems down the track.

Chapter Twelve
POSSIBLE HAIL

W hat if I told you you're more likely to have your server stolen or flooded than your data hacked? Many clients come to me concerned about the security that the cloud provides. They like to see their server sitting in their office, as this makes them think their data is secure. In truth there's a greater chance their server will be physically stolen or their premises will be broken into than their data being hacked into.

The data in cloud systems is stored in data centres. Whether private or public, these data centres are usually heavily protected by physical security. They'll have twenty-four-hour security staff on site managing the premises as well as other fail-safe measures.

Some of the data centres I've worked with are light airplane proof. This means they're designed to withstand a light aircraft flying into the building. With that kind of protection, you can seethe data centre at the building's core is safe and secure.

This is just one layer of the multilayered security systems that are put in place to prevent loss of data and data or hardware theft. These sorts of systems are just not affordable for small businesses. Small businesses, most of the time, need to focus on their core business. Adding in extra expenses for multi-level security systems dedicated to protecting their servers becomes very costly. Fortunately it is the cloud provider's responsibility to put in all these measures and ensure your data is safe. In turn small businesses can make use of these systems and security for the very low monthly fee.

Because security generally isn't tight at small businesses, we find it's easier for employees to steal data if they so wish. Often they can download folders or an entire database from a server because it isn't traceable. They can then resign and go to a competitor with a big chunk of that small business's intellectual property. However, with cloud providers everything is monitored, tracked and traced, which means every transaction that's done on the cloud-based system is accounted for.

I recently had a client who used Dropbox as their backup. Dropbox noticed that a large number of files had been deleted, so it sent me an email notifying me that a folder containing 1,000 files had been deleted. The email queried whether this was something I wanted or whether it was something I wasn't expecting.

I, of course, went to the client and told them that someone in their organisation had done this – intentionally or not – and I asked them what they would like me to do. They immediately had peace of mind knowing they could expect this sort of response with a cloud-based system. If this had happened on a simple server in their environment there would be no notification, tracking or tracing, and no one would know what had happened.

For many small businesses, data is their most important asset. Without it the business can fold and go under. When it comes to small businesses and data, having failover systems is imperative. Fortunately cloud-based systems put these failover systems in place.

They'll have the data stored on a specific server, which then has what's called mirrored drives. This means that if one drive fails they simply replace it and everything functions as normal. They'll have their own backup. They may even have their own data replication, which means that a data centre in one state is used in the unlikely event that a disaster occurs at their main data centre. In a main data centre, if something happens – whether it's a loss of internet connectivity or a catastrophic event – it simply cuts over to the data centre in the other state and all of the people using that particular cloud service are none the wiser. They just continue working as normal.

What if the cloud provider steals my data?

In most cases, your data is actually encrypted back on the server. This means that internal theft is impossible without the encryption keys. There will be different levels of security, meaning not everybody will have full access to the entire system. You'll need multiple levels to access the data. The cloud backup providers and good cloud backup providers will always provide the business responsible for the backups with an encryption key and that encryption key will be owned by the business itself. All data sent to the cloud backup company's servers is encrypted. This means any data stolen from those servers is completely useless.

Although there are risks with cloud-based systems, the risk is a lot lower than if you had the systems in house. At the same time, if you manage the data you're using in a cloud-based system to mitigate that risk – for example, downloading all your data once a week, month or day, depending on how risk averse you like to be – you'll always have access to that particular data.

Another security measure at the aforementioned light

airplane-proof data centre were airlocks. Visitors needed to be escorted through a security door to actually get into the building. The outsides of the buildings contained offices, which you must walk through after the airlock before you reach the data centres. Now you have to have additional access, or security, to allow you into the data centre. Inside the data centre are the servers that provide the services. They'll be database servers, email servers, file servers; everything that controls the cloud systems that you use.

In the data centre itself, you're required to accompany somebody who has a security pass or you must have one yourself. The security pass is to let you out. Once you scan into the data centre, you can't leave until you scan back out. This is a preventative measure in case of fire. In the event of a fire in the data centre, the room will fill with a gas, which removes oxygen from the air to suppress and put out the flames. Of course we wouldn't survive without oxygen.

It's imperative that these safety and security measures are put in place. They are put there for the end user's benefit, for the company, and for the small business, so that they can feel comfortable that their systems are secure and safe.

However, certain cloud providers don't put these measures in place. They may be a start-up cloud provider, so they don't have the capital. Going with a provider like this is far riskier than going with a larger company because they may not have the necessary security measures in place. They may be running their servers from an office, or – I shudder to think – from a spare room in a private household.

When choosing a cloud provider, it's imperative that you do your research. Make sure they have the right accreditations when it comes to security and data centres. Most of them will be very upfront with these.

AFTERWORD

C an you believe it? That's it! You've made it to the end of my book! I hope by now you realise that moving to a cloud-based system is not as daunting as it first seemed since, in actual fact, we've all been using the cloud in one form or another for a really long time… We simply did not call it the cloud.

Up until this point you may have felt as though the cloud or other IT systems were confusing, expensive and an unnecessary cost for small or medium-sized businesses. I believe that the cloud is for everyone. As you can no doubt tell, I am extremely passionate about the cloud and look forward to spending as much time with you as you need to get you confident in yourself and in your cloud-based system.

It may seem confusing at first, but if you are ready to take the step from having a server to moving to a cloud-based system, I would highly recommend it. Converting to the cloud allows small and medium-sized businesses to take advantage of larger business solutions for a fraction of the cost, saving time and money in the long term. I have seen so many people already who were hesitating and thinking about the cloud for a long time... but no one has regretted the switch. In fact, they wonder why it took them so long to switch in the first place.

I hope my book has helped you understand more about the endless IT possibilities out there and I hope you now finally understand what the fuck the cloud is.

Thank you for taking the time to read it all. It has been a pleasure for me to share this journey with you.

James

ABOUT THE AUTHOR

J ames is a communications specialist and systems expert, and is the Director of Right Click IT.

Born in Melbourne, James discovered his passion for technology from an early age. With technology seeming to flow through his veins, he was the go-to guy, helping his teachers fix computer problems and programming games.

His ability to quickly understand and simplify complex technological issues allows him to easily explain confusing concepts (such as the cloud) through his relaxed and approachable manner. James has a strong desire to solve any problem that is presented to him.

James worked for notable Fortune 500 companies including Hewlett Packard, Amcor, PaperlinX and Siemens after graduating from IT at Swinburne University. This time in large corporations allowed him to contribute to, as well as run some of the largest infrastructure changes ever implemented in those organisations.

A dedicated husband and father of two children, James most looks forward to his family holidays where he is able to see the excitement on his kid's faces when they experience new things for the first time. James enjoys anything that gets his heart pumping, including running, weightlifting, sky diving and fast cars.

What the F is the Cloud is his first book. This easy go-to teaches business owners to find a better way to systemise their processes and eliminate their IT woes.

WHAT THE F#*K
IS THE CLOUD
RESOURCES

Please contact James and the team at Right Click IT if you would like an analysis on the current IT systems in your business, assistance on setting up the cloud or online courses, please visit:

www.whatthefisthecloud.com

Right Click IT

Migration to the cloud and support

What's included	Business	Advanced	Enterprise
Office 365 Mailbox Options	✓	✓	✓
Domain Name Setup or Transfer	✓	✓	✓
Webhosting and Website Backup	✓	✓	✓
Unlimited Helpdesk Support 9am till 5pm	✓	✓	✓
Cloud Based Antivirus	✓	✓	✓
Access to Wholesale Hardware and Software	✓	✓	✓
DropBox Standard	✓	✓	✓
DropBox Advanced		✓	
DropBox Enterprise			✓
Access to Software Volume Licensing			✓
High Level Business Systems Advice and Consulting			✓

https://www.rightclickit.com.au/

www.ingramcontent.com/pod-product-compliance
Lightning Source LLC
Chambersburg PA
CBHW031900200326
41597CB00012B/499